The Blackbirds and Me
&
My Incredible Journey
with
The Lockheed SkunkWorks

By
Lew Williams
© 2019

Dedication

This story would not be possible without these people in my life:

My family – were and are the greatest supporters. They took the brunt of the times when I should have been at home, but wasn't.

Thanks to my late wife Pearlene and her strong resolve to maintain a family setting while I traveled.

Thanks to my children, Steve, Jeff, John and Regina who didn't always have their dad present because I was traveling for the job.

The last two years of my working career my thanks go to Nancy, my wonderful, second wife for her understanding of me and my love for another girl…. The SR-71.

And to
Jeannette Remak
Who encouraged me to write this story of my life.

And, most of all, I give praise to my God for allowing an ordinary man to become part of an extraordinary career.

Chapters

1. How I started my journey—4

2. Shock!!—11

3.A Day at the Fort—22

4. Off we go into the wild blue yonder--24

5. Life at the "Ranch"—27

6. Looking Forward—29

7. Overseas Deployment—33

8. Home—41

9. The end is near or maybe not—45

10. This is not the end—47

Appendices--48

(cover photo credit- Lockheed Martin)

Chapter I- How I started my journey

As a child, I developed a love for aircraft. I lived in Bellflower, California as a child moving there in 1942. There were three aircraft manufacturers in that Area; Douglas, Vultee and one more that I can't remember. Single engine, as well as multi
engine planes were always in the air around us. The other boys in the neighborhood and I could tell what kind of planes they were just by the sound of the engines.   So, my love for planes, you might say, started there.

Including Technical School for the Air Force, I spent four years in the Air Force from December of 1952 to December 1956. George A.F.B. was my one and only assignment while in the military.   I spent practically all of my enlistment in the Air Force trying not to get an overseas assignment. While I was in the Air Force, I had a couple of friends that worked in the orderly room. The orderly room was the place you went to if you needed something like a weekend pass, a leave, or a request for permission to get married, etc. My friends in that office gave me warnings about upcoming assignments so that I could try to get them or reject them.   I was always able to dodge these assignments and those men were a lot of help to me for that non- assignment. This office would be the office I had to submit a request to get married to my first wife Pearlene. The request would need the approval of the squadron commander: very happy he approved that request.

Prior to my discharge from the Air Force, a North American Aviation field technical group had been assigned to George A.F.B. to accomplish factory rework on the aircraft at the base where I was stationed. I became friends with the guys from North American Aviation and they talked me into going to work for them.

They told me that they would hand walk my paper work through the employment office and that all I would have to do is just walk in and tell them that I was there (I really had to laugh!). When I showed up to get hired, I had several names and phone numbers from these guys to help me through the process with the right contacts. As I spoke to the employment interviewer, we had a series of conversations that went like this: <u>Me</u>: "I was told you have an opening that fits me and that the pay would be a certain amount.' <u>Hiring Employee</u>: "That position is not available to someone off the street as you don't have the experience we need." <u>Me</u>: "Can I use your phone?" <u>Employee</u>: "What number do you want?" He then saw the list of contacts I had. I talked with my contacts one at a time as this employee watched and each time, my contact would say, "Let me speak to that employee." This happened several times including my request for the position as well as the entry level salary I wanted. Finally, this employee said, "Don't call anybody else! When would like take your physical and start the job? "

I did go to work with North American Aviation in late 1956, going through the field service technical school on the F-100 jet. Having never gone through a factory tech school on a specific aircraft, this was a real eye opener for me. I was taught many systems on this aircraft and some of them in great detail.

After I learned a system, I had to teach that system to others in the class. This was stressful as I was teaching my peer group in the class and they would be critiquing me after my presentation. I wanted to be good at what I learned. This job with North American, my very first job after the Air Force, sent me to England. Ironically, after I had worked so hard to not go overseas in the Air Force, here I was going overseas. My wife, Pearlene was pregnant with our second son (in those days the gender was an unknown) and her due date was in July. She had to resign from her long term job at Pacific Bell in North Hollywood due to her pregnancy. We lived very near her parents and they were on board to help her out as I would be gone.

After completing my factory training school, I was assigned to Wethersfield A. B. in England. After a few weeks there, I was sent to Nouasserur A.B. North Africa for transition training of the flight crews into the newly delivered F-100's. This is where I was when my son Jeff was born. He was three weeks old before I received the announcement of his birth. Talk about Snail Mail!! We did not have the electronic communication we have today. Oh how nice it would have been to have Facebook back then! This job was what made me realize that Field Service was where my heart was for my career.

In 1957, President Eisenhower cut the military budget and my job was eliminated. I returned home and worked for North American for a few weeks but the hand writing was on the wall. I found it better to jump to a new job rather than stay and wait for the inevitable layoff.

In late October 1958, there was no aviation work in California to be had. I moved my family to Buckeye, Arizona, where my dad had a café. He offered me a job and I took it. I was raised in a café and had done cooking all of my earlier life. My wife was not raised in a café and she never liked working there. She worked there just to make ends meet. Smashing hamburgers was not my desire for a career. I wanted back into the aviation field. I moved back to California, and decided to go to College and major in mechanical engineering. My thought was that with this degree I could branch out to any other field. However, I really wanted to get back into field service. I was living in Bellflower, California at this time. Bellflower is near Long Beach, California. At this time, with my wife and two sons to take care of, it was incumbent upon me to work while I went to college. Fortunately for me, Douglas Aircraft which was very near to where I lived and they were hiring.

I worked for Douglas prior to my enlisting in the Air Force in 1952. They offered me a job on the assembly line of the DC 8 commercial aircraft. This entailed working with the sheet metal structure of the landing gear doors. There was a team of four men doing this work. I was placed on the right main landing gear door. For some reason, the right side was always easier to fit to fair than the left side, or at least I thought so. Fitting door skins isn't as easy as one might think. It took some effort on our team's part to make an impression on the foreman. My team was able to do one and half aircraft in a shift which was very good. We were pushing the left side assemblers to stay up with us. It was a contest! I had requested the night shift at Douglas so that I could go to school in the day time. This would allow me to carry 12-15 units of credit. I could not carry this many units at night school. I enrolled in college and started work in the same week.

After being at Douglas in Long Beach for a year plus, I was laid off. This occurred in late 1960 just before Thanksgiving. I was drawing unemployment and I was also using my GI Bill for financial aid as I went to school. My family and I lived in an apartment on Main Street. With rent and utilities there was no other debt. We had enough money to get by on until Pearlene, my wife, could get a job with Pacific Bell her former employer. Pacific Bell would rehire former employees who had left with a good report, which she did.

We had made it through Thanksgiving, Christmas and New Years. In those days if you were on unemployment, you were required to report where you had been to check for work. I knew who was and was not hiring. I also was not required to go out of a 25 mile radius to look for work.

Money was getting very tight one week and my unemployment check would not arrive for a couple of days. It was one of those times when I didn't have two nickels to rub together. We ran out of toilet paper. You know that is major! Pearlene started crying and was upset about our situation. I did some quick thinking, and right away I had a perfect plan. I was lucky enough to have achieved the wonderful status of owning a Chevron Gas credit card. I would take her to see her mother who lived in San Fernando. This was about 35 to 40 miles from where we lived. On the way to San Fernando, I decided I would stop by Lockheed Aviation in Burbank to see if they were hiring. I loaded the car with gas and my family; Pearlene and sons Steve and Jeff. Off I went to make an attempt to look for work thinking that Lockheed was not hiring. I'm ashamed to say this but I was only looking for work to make my wife happy. I was quite content with unemployment and my G.I. Bill. I was also hoping my wife would go back to work at Pacific Bell and I could take my leisure going to school.

We drove to Burbank and I pulled my car (1957 Chevy Bel Air) right up in front of the hiring hall at Lockheed. There was not another car in sight! I was right, THEY WERE NOT HIRING! I had my resume in hand and entered the hiring hall. For some reason, the required information on a resume included the last ten years of your life: addresses, jobs held, types of security clearances, schools attended, etc. I had moved around to several different places and had employment or school with each move so there was a lot of information to remember.

As I sauntered into the hiring hall, sure enough there was only one other soul in that huge room. He was setting behind a counter with his feet resting on the counter. I approached him, gave him my resume and said, "You don't have anything, do you?" After reading several pages of my resume, he said, "I think I do." Well, my heart sank; I really didn't want to go to work. As I said previously, I wanted to go to school at my leisure. My whole being sort of collapsed and I didn't have the resolve to walk out right then and there. No, instead I followed his instructions to another area.

He told me to follow the yellow stripe that was painted on the floor and talk to the lady behind the desk. I walked several feet to the lady. We spoke for a little while and I told her that I had my wife and children in a car parked outside and it was getting close to lunch. I told her that I needed to get them to a place where they could eat and be taken care of. She assured me that this wouldn't take long, but could I please go to Building 82, knock on the door and wait for someone to answer the knock. Someone finally answered and I met Dick Greer who was a department head. We went to a small office where he asked me question after question. After a while, he told me to go back to the lady I had first talked with. This lady made the agreement for amount of pay and details for my security clearance.

We finally drove on to see Pearlene's mother. The whole episode made Pearlene happy, but I was dumb founded! I had just been hired and I didn't want a job. Drat!!!! It was a sour note that I had been hired and little did I know that I had been hired into one of the most prestigious divisions in Aviation. As I look back on this experience with my lack of enthusiasm for getting a job, I am astonished that I was even considered for such a great job!

I returned to Lockheed the next day and started the process of filling out the paper work; getting my identification card, badges, and physical exam. The long arduous task of filling out the security clearance had started and I later learned that this was a CIA security form! As I said previously, I had to go back ten years with names and addresses and the list went on. I was told to report to Building 82 on Monday morning for work.

I would like to say at this time that on my first day of work, upon arriving home, Pearlene greeted me at the door and asked about my day. She then said, "Oh by the way, I had a call from Pacific Bell today telling me I could go to work." She said, "I told them I don't need a job anymore, my husband got a job". "Thank you very much for calling." So much for my leisurely plan to go to college!

Chapter II- Shock!

It was February 1961. I walked into the lane that led me into the entrance of Building 82 in Burbank; it was my first day at work with Lockheed and ADP (Advanced Development Projects) which didn't mean a thing to me at the time. As I said previously, I did not realize what a prestigious program I had been hired into. I stepped across the opening of the big roller door into an experience that was the most remarkable of my life. I waited in a line with about four or five men that had just been hired. We were very unsure of ourselves. We had just entered into the "World of Make Believe!" I was escorted to my supervisor and he gave me a tour of just which areas I could be in. The building was compartmentalized into small areas and you had to have prepositioned clearance that was designated by the type of badge that you had. During this tour I saw **HER,** the *Article* also known as the A-12. She was the most futuristic thing that I had ever seen!

 I didn't know if it was an aircraft or space vehicle! Have you ever had a dream where you were involved in an exciting project the likes of which you could have never imagined being involved with? That was how I felt on that day in 1961!

I felt as though I was standing in front and beside what looked like a very real space ship. I had seen many Buck Rogers comic books as a kid. I always wondered if those comic book aircraft could fly. Now I was wondering would I ever see this beauty, the A-12 fly or would I be denied that pleasure? However, just to be able to know that I was going to be part of this incredible journey sent chills up my spine. We returned to the area where I would be working and I was given a briefing as to what was going to be expected of me. The requirements were way above my previous jobs. Supervisor Dick Hough (pronounced Huff) told us that he was aware we all were entering areas beyond any present experience and we would be learning as we went along. What a relief!!

My assignment was to the research and development hydraulic laboratory. Here was the heart of the hydraulic system for the A-12 vehicle. Along with the hydraulic system, I would be working on other systems of the aircraft as well. It was a real learning time for me and the two other men; Johnny Holms and Ben (last name unknown). We had been brought into the program and worked in the same lab. We had just been approved for a CIA security clearance.

We were told that if we lost our badges, we would lose our jobs. The reason for that was if that badge was lost or stolen, Lockheed would have to re-issue new badges to all of the employees on that program. There was also the proviso that should we have to go a dentist and be put under anesthesia or any kind of medical procedure, we would have to have a security escort in the room with us to make sure that we did not reveal any of the classified information we were privy to. To my knowledge this never happened but at least the guide lines were in place if it ever did.

I had the privilege of working at the Skunk Works while Kelly Johnson was there as VP. Kelly was the master designer of aircraft for Lockheed. This I learned was the epitome of excellence in the world of aircraft design. The men I was surrounded with at the Skunk Works were the best of the best under Kelly's direction. Now, I was a part of that Advanced Development Project.

As I stated earlier, the first time I saw the "Article", she was a sight to behold! She was so futuristic with that long stiletto shape. The size of her was just such a departure from the aircraft that I worked on before. I mean, this was the 1960s, but this vehicle looked like something out of the future! And it was. In my position, I had access to the Article on a daily basis, and had hands on type activities working on her. We would work in the Lab testing the hydraulic valves, tubing, and seals. Our shop was next to where they were doing silver brazing of the hydraulic line joints. There was a procedure that was required where the personnel (these guys wore white smocks neckties and dress shoes. I thought for sure they were engineers but I never really asked) had to evacuate the outside air from around the joint which had been placed into a glass cylinder. They then vacuumed any oxygen out of the cylinder and then filled it with gaseous nitrogen so that the braze joint would not be contaminated. It was really high tech stuff for that time.

We were also required to pressure test all of the tubing to "proof pressure test" of 5000 psi before they could be installed in the aircraft. Once the tubing was installed by the plumbing department, we (the hydraulic lab) had to leak test the system to make sure that none of the joints leaked. Johnny Holms, Ben and I pumped the very first blood (or hydro fluid) through her systems. I have always considered this to be a real milestone in my career. The oil that we used was so expensive that we had a cheaper oil to use for check out. After the aircraft hydraulic system was cleared for operation, we then flushed out the systems and replaced the test oil with the flight oil. Our oil was so sensitive to oxygen that we had to keep gaseous nitrogen as head pressure so that oxygen could not get into the system and create a contamination problem.

We had a cart where we stored the hydraulic oil. This cart had a motor inside that we could circulate the oil. It also had a vacuum pump which vacuumed out the oxygen from the top of the unit. We then ran nitrogen from the bottom of the tank in order to have oxygen free oil for the aircraft. We would run this cart for twenty four hours and then check to see if the oxygen level was acceptable. If this wasn't acceptable, more vacuuming occurred. If there was any oxygen in the oil and flown in the aircraft at the extreme temperature that she raised, it would cause the oxygen to burn and cause a varnish to form on the very small spools in the transfer valves. This could lead to flight control as well as other systems problems.

In addition to the testing of the plumbing and seals and the valves, we also refurbished a device called a hydraulic mule/gig. This was a portable electric motor driven unit that supplied hydraulic power to the vehicle on the ground so that we could simulate operational functions. The gigs had a two system capability that would allow oil to be supplied into two systems at once. We were using existing Air Force equipment as our gig and therefore it came with "RED 5606" hydraulic oil; standard for other aircraft at that time. In order to keep our oil from being contaminated, we had to disassemble the gigs and flush every component clean of any 5606 oil. This took on a huge amount of work. We were very busy disassembling gigs, testing hydraulic lines, cooking hydraulic valves at working temperature. This modification was very time consuming and after we had completed three Gigs' the modification was turned over to another crew somewhere outside our location. We were happy to pass this project on to someone else!

As the vehicle started to take on a completion state, the hydraulic lab would operate the hydraulic gigs to do functional tests which included the flight controls, landing gears, and the inlet control system.

The first inlet control system used was a Bendix inlet control system. Evidently there were some problems with this system because as aircraft 128, 129 and 130 came out of assembly they came out with a new Lockheed Inlet Control System.

I personally had been able to work with the flight control system and with the engineers that had designed the servos that port hydraulic fluid to actuators that move the flight controls. When it came time to do the functional test I was very pleased to be a part of that team. I would be the operator of the gig and I became aware of the frequency response test. This test was a requirement whenever a flight control servo was replaced. The test was also required if there was a flight write up that might indicate the flight controls were having a problem.

The flight controls, inlet system and the landing gear were the main hydraulic systems of the aircraft. We also preformed checks on the Gaseous Oxygen system (later up graded to LOX, liquid oxygen), the Cryogenic LN2 (liquid nitrogen) system, and all of the pyrotechnical units that were used in the ejection seat of the A-12. As I remember, we had to do some modifications to a seat called the Lockheed ejection seat. This was a very interesting job as the seat structure was very complex. The shoe spurs had a cable cutter that we tested for operation to make sure that the anvil and cutter were installed correctly and could move under the air pressure required.

Many times the engineer in charge of a particular system would bring us a new valve and simply say, "make it break or take it to failure!" This always delighted us because it gave us an opportunity to see if the design could hold up to what it was designed for. This brought out the younger boy in us: boys like to break things!

Some examples of taking these parts to failure included building new items to check them out. For instance, if the item was going to be in an enclosed environment and at temperatures over 500 degrees, then we would build a special box to put in an oven to test the unit.

In another test, we had an electric motor that drove a cam so that we could simulate vibration. This unit was hooked up to a block of hydraulic tubes that were pressurized then placed into the oven and run to a million cycle to check for unit failure in the joints.

We had big ovens that went above 500 degrees F where we could simulate flight temps that would exceed the normal operating temperatures environment of the unit. We had many opportunities to complete the functional test of all the moving surfaces of the aircraft.

A-12 under construction after arriving at Area 51- fuel and hydraulics being checked   (Phoenix Aviation Research-Roadrunners Internationale)

Fuel testing for the A-12 at the Fort
(Phoenix Aviation Research –Roadrunners Internationale)

The fuel test rig located at the Fort (Phoenix Aviation Research –Roadrunners
Internationale)

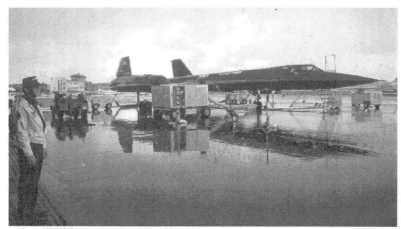
Lew Williams overlooking his charge , the SR-71 at Farnborough UK
(Leland Haynes Collection)

Lew Williams with his crew at the Farnborough UK Hangar with
dignitaries looking on. (Lealand Haynes Collection)

The cockpit of the A-12 in the oven being tested for heat absorption testing for the A-12 at the Fort (Phoenix Aviation Research-Roadrunners international)

Rear Row L-R: ??, ??, ??, On Ladder ??Front Row L-R: ??, ??, Wally Baxter, Fritz Frye

500th flight of the A-12 at Area 51` Fritz Frye at the right hand corner. (Phoenix Aviation Research –Roadrunners Internationale)

The A-12 being readied for transport. The rig was built at the Lockheed Burbank plant(Phoenix Aviation Research Collection)

The A-12 sits at the roadside in her trailer for a break in the long trip to the Area51 "Ranch" (Phoenix Aviation Research Collection)

Chapter III - A Day at the Fort.

In addition to our shop activities, we would have to go out to a place called, the Fort. In this place, we used highly dangerous liquids and explosives. The place had outside walls made of corrugated metal panels and chain link fencing so no one could see what we were doing. This is where we tested the flow rate of the cryogenic LN2 valves for flow rates required. Here we also tested the explosive devises that were used to launch the missiles from the YF- 12s mission bays.

One of the first things that I saw when entering the Fort was a mockup of the cockpit of the aircraft on a frame. The frame was set upon something that looked like railroad tracks. These tracks allowed the mockup of the cockpit to move in and out of a very large oven. I was never around when they were working with that system, but it was very impressive. I remember meeting Fritz Frye there who was the supervisor. Lots of other impressive things were in the making there. However, my clearance didn't allow me the privilege of more information.

I worked for two and a half years in Building 82. Then, we were moved to buildings 309 & 310 where we started production of the A-12. After we completed the assembly of the A-12 Aircraft and had run all of the necessary tests to make sure it was ready for shipment, it was transferred to AREA 51. AREA 51 was also known among the Lockheed employees as "The Ranch". This was a lonely place in the Nevada desert and was a very secret place.

Moving the aircraft to AREA 51 was accomplished by removing the outboard wings, rudders and then placing the rest of the aircraft on a flatbed truck. All of us in the hydraulic lab had to drain all of the fluids from the systems and clear the lines of any other fluids. This part of the job was very exciting as we were about to finally deliver our product to the flight test location.

I worked for two and a half years with stimulating work that challenged me and taught me so many things. I was fortunate to work with some very knowledgeable men; engineers and designers: Tom Takesue/ hydraulics, Tony Faseotti/landing gear and Al Foggner controls and servos. There were others who did a very important job of designing seals that would withstand temperatures and would seal surfaces properly so that hydraulic fluid would not leak where it was not supposed to leak. All of these people had great knowledge in their field and it was a great learning experience for me.

After working two and a half years at the Lockheed Burbank location and having learned the hydraulic system as well as other systems, I had a great desire to go into the field where the vehicle was going to be flight tested. A man named Frank Berttelle was in charge of supplying the man power required for AREA 51. This was a top secret test site used by Kelly Johnson and the government. I put in a request for transfer to the AREA and after a month had passed, I hadn't heard anything about the job. I sought out Frank Berttelle and had a meeting with him. He informed me that he no longer was responsible for the man power at the AREA but he pointed out a man sitting across from him who was now in charge of supplying that man power. This man asked me many questions about the A-12 and after a lengthy discussion, he said, "I just don't think you have the qualifications for this job. I was furious! The people he was sending to the AREA had been trained by our team. I really exploded and I informed this man that I was experienced in the hydraulics, fuel, seat pyrotechnic and a few more items that I thought were important. Frank Berttelle spoke up and told this man that he would soon be needing a person with these systems specialty training and inferred it might be wise to hire me.

Chapter IV –Off we go into the wild blue yonder….

I asked for a transfer to the AREA to work on the A-12 during flight test. After a few months of delay, I finally received a transfer to the AREA in December of 1963. I was now officially going to work at the AREA. I was told that I would be required to report to work at the shack, just opposite of the gate where parking was available. This was at the Lockheed Plant flight line in Burbank California. This was my first experience with unmarked flights that took us back and forth to the AREA.

At that time, we only had one family car. Pearlene drove me to work along with the boys so I could start work at 5:30 in the morning. We said good-bye and we were to meet again on Friday at 4:30 pm when she would pick me up. I proceeded to the room where check in was done and reported in. I was then given a loading number.

We were then transported to the AREA by three Lockheed Constellation aircraft. (Connies, as they were nicknamed). The flight crews that flew the Connies were veteran pilots and very experienced at what they did. I remember on one occasion we were on the takeoff roll and it was very windy that morning. We were almost at lift off when I saw, out of my side window, a very large panel from what I believed to be one of the hangars roofs. It blew on a path right at us on the runway. I was sure we would be hit! The pilots did some quick maneuvers and that Connie did a dance that day the likes of which I have never seen again! Later upon arrival to AREA 51, totally unscathed, we all had a great time shaking the pilot's hand and thanking him for his quick actions. This was the only problem I ever saw with the Connies when flying to the AREA.

Prior to going to AREA 51 the first time, I helped load Article #129 aircraft on a flatbed to send it to the AREA from Burbank. This would be the plane I would be assigned to after I arrived there two months later. When I arrived for my first work day at the AREA, I was greatly impressed with how stream lined their operation was in this area. They were so efficient with our lodging. We had a wood structure house with four bed rooms with two men per room and a common area in the middle of the building. We had maid service and all our meals were provided. My roomy was on night shift and we very rarely ever saw each other. I think this was standard procedure.

I was placed on the day shift prior to coming to the AREA. My hours were 6:30 a.m. to 4:30 p.m. It was normal to work 10 to 12 hours a day , five days a week. We soon found that week end work was always available if one wanted to work. Beside my work on Article #129, with my experience, I worked on other aircraft when extra help was needed.

I had a very extensive working knowledge of the hydraulic system and its components and when there was a flight control problem, an engineering staff from Burbank would come to the AREA. I would operate the hydraulic gigs for them.

I worked closely with them and the rigging crew. Many times we would have to replace a flight control servo which necessitated breaking into the hydraulic system and replacing the fluid when we were finished with the rigging. I also had the opportunity to work in "Burbank with the designer of the first hot gig, Howard Choennor. This GIG would heat the fluid to a 500 degree temperature and we would soak the complete aircraft hydraulic system until 500 degrees was accomplished to make sure these components worked well under high temperature in flight and did not leak.

I saw some flights of the YF-12 for my first time at AREA 51. This was the follow on of the A-12 that I had previously worked on. I also saw the M-21 and the D-21. The M-21 was as a reconfigured A-12 that carried the D-21 drone on her upper fuselage. They were housed in TOP SECRET ACCESS ONLY hangers. However, when they rolled them out for a launch, they were right there by our hanger. Because of this, we were able to observe their pre-flight procedures. They had a few different steps but for the most part, it was the same as we used on the A-12. They did have a different ANS (ASTRO NAVAGATION SYSTEM) alignment procedure. It appeared to be less time consuming than ours. Often there were opportunities for overtime which allowed me to work over the weekend. One weekend I worked on that secret project. It really didn't require anything I wasn't already doing on the A-12 #129 but allowed me to be exposed to new systems.

I was on the ramp fairly regularly with the launch of our A-12 #129; The YF-12 would be there with us before she was transferred to Edwards AFB. She was sheltered in a hanger behind us. When they brought her out for launch, you just couldn't ignore her.

Chapter V—Life at the "Ranch"

One of the real treats at The Ranch was the Chow Hall. I am not kidding!! For such a large operation, it provided us with great meals! For breakfast we had eggs cooked to order, any and all kinds of breakfast meats. I mean all of the foods that were normally used in breakfast like sausages, bacon, etc. You must remember, I came out of a family Café business, and the presentation of the food was of an interest to me. Every evening they had there own special items. I do remember, every Thursday night, it was steak night. They served the largest steaks that I had ever seen! They were cooked in varying stages, rare, medium, medium rare, whatever you wished, and as many as you could eat! Another night, we were served a "Steamship Round Roast". This cut is traditionally used for the roast beef at a buffet. This was the biggest roast beef I had ever seen this side of the Café business!!

There were tables that seated six or more men. This was one place where I met new people. Even though we saw each other at our work stations, we never had a chance to talk to each other until we sat at these tables together. One man I met at this location was Colonel Robert (Silver Fox) Stevens. He was the first military pilot of the YF-12. I had bumped into him at the Chow Hall and we sat together for lunch. He was such a nice guy. I still had no clue as to who he was at this time.

In addition to the Chow Hall, we had clubs for recreation. I was in a tape recording club where we could copy master tapes of music. I bought a reel to reel recorder and started to make my own music library. We had all sorts of music to copy. The men would share their tapes freely. I also heard of clubs like rock hunters, they would check out a vehicle from the motor pool and go out exploring the desert for rocks. Another club I joined was a

---

27

Bowling league where we bowled on Wednesday night. I had an opportunity to meet people that I didn't see at my normal work location. I met some of the Air Force personnel. I also met men from the David Clark Company that manufactured the flight suits for the pilots and some other men from the north end of the AREA who were with the Lockheed Air Service out of Ontario, California. I always looked forward to Wednesday Bowling Night. I also joined a Bible study group.

Chapter VI—Looking Forward

In 1965, the CIA's A-12 was getting ready to go operational. Their project name was "BLACKSHIELD and the aircraft was called "OXCART". She was also known as Cygnus, a name given to her by her pilots. Frank Bertelle was in charge of this operation and I asked to be a part of this program. I was a little slow in putting my name in for the transfer and was told the operation was full. A few months later, I was made aware of the soon to be delivered SR-71 owned by the Air Force. With this news, I asked Frank Bertelle for a chance at a field service position with the SR-71. He said, "Have you any special talents with the aircraft?" I had the feeling that Mr. Bertelle was teasing me because of my *rant* in asking for a transfer to the AREA. I came back at him with all of my experience and anything else I could think of. Mr. Bertelle said, "ALRIGHT, ALRIGHT ALREADY!!" I will give your name to a man in Burbank who is hiring men for the job". I thanked him and with a twinkle in my eye I said, " I do have all those talents I spoke of." I went on working at the AREA for what seemed like a very long time and had not heard anything about the position with the SR-71. I happened to bump into Mr. Bertelle at work again and asked if he had heard anything of my acceptance into the Field Service of the SR-71. He quickly said, "no" but he gave me the name of Ed Packard. He told me this was the man I should talk to at Burbank if I wanted the job. I arranged a meeting with Ed on a Friday. If I could get the very first flight out of the AREA, I could make the meeting with him on time. I was able to make that flight and I met Ed in Building 311 (the engineering building) at Lockheed Burbank. We went into a room just off the main entrance. I asked if I had been selected for this position and if not, what would I need to do to get the job?. Ed told me that I was still on the list for the position but the requirement for more people right now was not happening. I was relieved to hear I was still on the list but impatient to be hired. He told me to wait and when the opening happened, I would be called up.

At one time I told my wife Pearlene that I was applying for a different job that might move us to Northern California. She desired to live in Northern California with the caveat that she never wanted to live in Marysville because that was the area that her family friends had experienced flooding.

In June of 1966 I was called up. I was told that I would be moving to Northern California and reporting into work at Beale A.F.B. on the SR-71. Wouldn't you know it, the nearest town to Beale A.F.B. was Marysville. Pearlene was not excited about this move but came along anyway. There were some additions to the family before we moved to Marysville. Our third son John was born in December 1962 and daughter Regina in July 1964. When we moved to Marysville, there were six in our family.

When I reported to work, I checked in at Building 678 at Beale AFB. This building was up on the hill away from the flight line. This building housed most of the technical representatives for the SR-71 Program. I was given directions to the flight line office and promptly met a couple of old friends, Bill Brown and Tony Chiappetta. I knew these men from the AREA. I also met Jack Boen and two other friends, Andy Stumpf and Larry Bassett at Beale A.F.B. This began a twenty year friendship with all these men. In the beginning, I worked as an APG (air plane general) as did these other men. We each had responsibility for three SR-71 aircraft that were already at Beale. We would be with the crew chief for the everyday maintenance of the aircraft and for any other needs. We had other men in the office at Building 678. These men were Paul Mellinger (manager for the SR group), Jim Cook, John McCann, Leon Crook, Ken Nudson, Bob Carter, Aaron Higley and Jack Johnston. Our secretary was Margaret Martin who would, much later, become Mrs. Paul Mellinger.

I have to interject this about the caliber of men that I had enjoyed working with. When you talk about mentors in your life, how many can you really say were "mentors"? Through my career at the Skunk Works, I had been rubbing elbows with men of great wisdom in their field. The same thing happened to me at Beale AFB. I often described myself as a small oak tree among the Giant Sequoias! Aaron Higley was the most influential mentor that I had. Aaron took me under his wing and had me in on every work situation that he became involved with. My understanding of the SR-71 was greatly enhanced because of him. Jim Cook, John McCann and Paul Mellinger were some others that invested their wisdom into me. Aaron was the person that covered the inlet system and the auto flight controls. These systems were demanding more coverage than Aaron could cover. In order to aid Aaron because of his over load, I was asked to work with him. I told him that I was familiar with the frequency response in that I had run the hydro gigs for that test at the AREA. After several weeks Aaron asked that I be placed with him to be his second.

I received the upgrade position. I was used to de-brief flight crews concerning any in flight maintenance issues. If there were maintenance issues required after a flight, I would brief the next flight crew (Pilot and RSO) of any maintenance that had been accomplished.

In addition to de-briefing and briefing flight crews, I was working with several other shops. For instance, if a flight control servo mal functioned and needed to be changed, it took both the hydraulic shop and the SAS (Stability Augmentation System) shop to facilitate this repair.

The SAS shop which was the automatic flight control shop would do a frequency response check to see if the servo would meet the specs required for service. If the servo did not past inspection, the hydraulic crew would change the servo. These teams working together sometimes required odd work hours and long hours. They would constantly call Aaron and I to come out to assess the problem and this call always seemed to often come after mid-night. We would spend several hours there before we could determine if the servo was functional or not. All of these shops working together was vital for good functional flights.

There were occasions when the aircraft would have to declare an emergency and land away from home. These occasions would set into motion a recovery situation where a team of maintenance guys would gather up all of the necessary equipment to; tow the aircraft, refuel the aircraft, set up the start cart so that we could start the engines and a hydraulic gig along with any other necessary equipment that was on their check list. This recovery gathering took several hours but we had already been in voice contact with the flight crew. They gave us some idea of the problem so we knew what spare parts to bring along.

Also, on many occasions, I would have to call home and ask Pearlene to pack me a suit case of clothing for 3 to 5 days. If possible she would bring them to me at the base. I would not be able to tell her where I was going or when I would return.

Pearlene was always very responsive to these requests and never complained about my disappearing for many weekends with no warning. The wives of this unit were all special as we never had a divorce in the whole group.

Chapter VII- Overseas Deployment.

We had been at Beale for almost two years when we were alerted that we would be going to Okinawa, Japan for extended duty. Our routine of daily operations was about to be put to work at real targets. This information was secret and was not to be made public. We could not even tell our families where we were going or when we would return. We were only able to give them the office phone number and if required, the office would relay any information the officials felt necessary. This went on for years. Pearlene found out more information about my job while she was at the beauty shop with military wives. Sometimes this information was so close to the truth, about our secret mission that I knew some military men had loose lips!

The rotation teams were set up and dates of deployment were recorded. We started to place into motion the necessary preparation to have this accomplished on schedule. The ad von group of maintenance men and officer core deployed on Feb. 29th, Leap year Day in 1968. They traveled aboard several C-141cargo aircraft and some KC-135 tankers.

Detachment #1, Kadena A.F.B, Okinawa, Japan was the operating unit that we flew out of for over twenty years. We replaced the CIA group that was there with the A-12s. We first started out with three SR-71s on site and the necessary flight crews and ground crews to operate all of them. This included six Lockheed Representatives. I was not supposed to go overseas until the fourth rotation. However, for some unknown reason that was changed and I was in the second rotation group. I departed in May of 1968 for three months and at that time, we flew in the KC 135s with all of the other military personnel. The KC-135 did not have a comfortable seating arrangement and on a 16 to 18 hour flight, it was a hard trip. One of the things that we looked forward to after this long trip was a visit to FBIS ("FEDERAL BROADCASTING and INFORMATION SERVICE"). That name was used in the early days. It was later renamed and called "Skoshi

Koom". This was one of the best restaurants on Okinawa. This was a favorite place for the crew members to gather for eating and drinking and gambling. I was able to have many a wonderful meal there.

On my first arrival to Okinawa, I was housed in a hotel not far from the base. It was not much of a hotel and so noisy at night that it made it difficult to sleep. We were eager to find another place to stay. Bill Brown and I went out and found another hotel called, Iha Castle. For the same price we could stay in a four star hotel rather than a two star. We all moved there and stayed there for a couple of years.

Our daily work schedule was from 7:00 am until 8:00 pm, 7 days a week. When the SR-71 flew, we would collect the MRS (Mission Recorder System) and analyze them. This system is known as the "Black Box' on commercial aircraft. We would report our findings to the shops concerned, who also looked at the same tapes. When they found problems they would report them in the flight log of the aircraft to generate a work order.

Our teams worked hard together and on slack times, we would have fun recreation times together. These fun times consisted of BBQ's with chicken, burgers, dogs and of course, beer by the case.

We had many flights that flew close to the DMZ as well as Korea. We had alternate bases in preselected areas should a flight crew have to abort. We would have some recovery equipment already on site for these repairs and sometimes we had to bring equipment in from our home base for the repairs. We hauled the equipment on the KC 135, repaired the aircraft as quickly as possible and returned her to her home base so she would be ready for another mission.

I remember my first recovery overseas. We had a plane abort into Utapow, Thailand. We had received word that the plane was going to land away from home and we started the procedures loading the KC 135. We picked the replacement parts needed as much as

Could be determined from our brief conversations with the flight crew. I was told that the hydraulic system was in need of repair. Off our team went with one well packed tanker, an eager crew and one reluctant tech rep. I was reluctant because I had been working all day and was not looking forward to this long flight.

We had been airborne for a few hours when we were told that we were crossing the DMZ. The KC-135 tanker has no windows to look out of except for two small windows in the emergency exit hatches. I was looking out of one of those windows as we crossed the DMZ and I saw what looked like flash bulbs going off. I mentioned it to the others on board that they were taking pictures of us. Very dryly they said, "Lew, those are not flash bulbs. They are shooting at us!!" I went to my seat, hoping the tanker could go faster! Boy did I feel dumb!!!!

We would arrive at the recovery location and debrief the flight crew and assess the problem without benefit of the MRS. It was not a bad situation but it really required comprehension of the systems and aircraft to get her repaired. Once we determined what was broken, we would facilitate the repair. We always seemed to bring all the correct items needed for repair. DANG WE WERE GOOD!!

It was always a thrill to get the aircraft ready for launch. Somehow, the people from the host base, having never seen the SR-71 before, would somehow find out our take off time and start lining up alongside the runway just to get a look as she took off. Our flight crews did not disappoint any of them. A max performance take off was one that we as a recovery team, thrilled to as well. She was a beautiful sight!

After our recovery was over and the airplane had been launched to go back to our home base, we boarded a KC-135 and slept most of the long flight home. We had spent all night and all day getting our girl ready to return.

Some other occurrences we had with the SR-71 aircraft included one aircraft which crashed on landing wiping out the under-carriage. It was not repairable. We also had one aircraft that did a pitch up just after disconnecting from the air refueling tanker and went down over Thailand.

We had one SR lose an engine fuel line and caught fire at DET 1. We needed to change the out board wing of this plane. Civilian mechanics from Lockheed Palmdale brought the necessary outboard wing to make this repair. The military mechanics were available to assist them in the repair as needed. After this repair was completed, we were required to do a Functional Test Flight which required a Flight Test Pilot to complete the flight. Col. Tom Pugh, a military test pilot from Palmdale, performed this duty. After many tests on the ground and much discussion, we were ready for the test flight. Naturally, several sub-sonic flights would have to be performed before we could take her out to speed. After accomplishing the required tests, we were able to prepare for a flight back to California for further examination. This aircraft returned home to the States at a mere Mach 2 instead of Mach 3 and still at high altitude.

Typhoons were a constant threat during the typhoon season and our hangers were built to withstand winds greater that150 miles per hour. The normal wind speed for a typhoon was 110 to 120 m.p.h. We had a skeleton crew stay with the aircraft during this time in case any damage came to the hangar and needed immediate repairs to protect the airplane.

When we first went to Okinawa, we were allowed to talk to our families via Ham Radio. The DET fixed us up with a powerful Ham radio station. In our ranks ,we had several licensed Ham operators who volunteered their time and helped us to stay in touch with our families.

As I mentioned before, most of our civilian travel to and from the DET (to Okinawa with return to Beale) happened on a KC-135. The Air Force had an incident on a KC 135 losing cabin pressure with a civilian on board. The Air Force re-thought their position on allowing civilians to fly overseas on a non-military personnel flight. We started to fly out of Travis Air Base on Military Air Transport, which had proper seating and accommodated passengers. We did this for a couple of years. We were told that the Air Force was charging the Lockheed Company the same amount of money to fly with them as it would cost to fly commercially. So, the company started allowing us to schedule our own flights. This type of travel continued until we closed the overseas DET in January 1990. I enjoyed scheduling my own non-military flights so I was not subject to military schedule and delays.

In 1974 the decision was made to go public with the SR-71 and arrangements were made to participate in the Farnborough Air Show just out side of London, England. It was a real privilege to be selected to travel with the ground crew for this event. Thanks to Paul Mellinger, I was selected to make this trip. I started the briefing meetings and preparation work for the trip..

The target date for the Recovery to be on site in England was August 27. 1974. We needed to be on site prior to the attempt of the SR-71 establishing a world speed record from New York to London. As we drew closer to the launch date, we wanted to travel in more isolated areas so as not to disturb populated areas with our sonic booms that we were known for. We moved our flight and acceleration patterns around so that the same people didn't experience them repeatedly. To have the surveillance like the SR-71 was able to provide, it comes with a price. The sonic boom was the price tag.

The ad von crew was in place at Farnborough Royal Air Field in England on September 1·1974. I was up at 4:00 a.m. on this memorable day to be at the air field along with Paul Mellinger and Joe Gully our Lockheed photographer. We wanted to be there early so we did not miss one moment of this exciting day.

We received word from the Air Force crew members that the SR-71 was in the area and would be landing soon. As the maintenance personnel started to pre-position the recovery equipment, our excitement started to build. We began a watch of the skyline searching for the familiar profile of the SR-71. It wasn't long before we saw her and we were not disappointed. She came in first in a high speed pass and a climb out that was outstanding! Then, she came back over the field in a slow flight so that everyone could get a good look at her.

After the SR-71 landed, we positioned the crew stands for deplaning and get the official record. The monitors, people from the official recording office, were up on the stands to confirm the crew was who they were supposed to be. Next, the maintenance crew was allowed to help the flight crew deplane. After this, the mob of Air Force dignitaries as well as a Senator and other Government officials came and surrounded the crew. We delayed our ship side chat that we normally had to allow these dignitaries access to the crew. After a while, we were finally able to gather the crew in a quiet area so we could de-brief them and find out if there were any in flight problems that required repair. We had a couple of squawks but nothing of consequence. A normal post flight inspection was completed and the crew was released to change their clothes. It was a relief for the flight crew to get out of these cumbersome suits.

Everyone in our group stayed in a hotel in downtown London called the" Churchill Hotel". My son Steve arrived from Germany to join me at the Air Show. He stayed in the room with me and came with me to the Air Field on a daily basis. He took over my hand held camera and shot many pictures. I was ever so glad Steve took pictures for me as I would never have gotten a chance to do it myself. After the air show was over, Steve had to return back to his Army duties in Germany. I was so happy his assignment in Germany aligned with my assignment in England making easy access to be together.

On September 1st at 10:00 pm, Steve, Paul, Joe and I were trying to go into a restaurant to get something to eat as we had not eaten since noon. The reason we were so late in getting dinner was that we had to have the film developed that Joe had taken that day of all the dignitaries, pilots, the arrival of the aircraft, etc. We needed to get this film on a flight back to Los Angeles as soon as possible. This film was classified and we had no official courier to handle the package. Ann Widdefield, wife of the RSO (back seater in the SR-71) was scheduled to go back to the U.S. the next day. We asked her to carry the film to LAX. We told her she would meet a man in a trench coat there and he would ask for her by name; something that seemed it was taken right out of a spy novel! Well, she carried the film back but did not enjoy her flight. She was very frightened by this responsibility. The film was delivered on time to the Lockheed courier and the announcement was made that day about the SR-71 speed run.

The restaurant we chose for dinner was closing it's doors as we arrived. We were turned away and we started to look for other options. As I stated before, I had been up since 4:00 a.m. I had my contact lenses in and they were giving me a problem. It was my habit to remove them when they began to irritate my eyes. I always carried my contact case for such a problem. There I was on the cobblestone sidewalks of London at 10 p.m. with little light. I started to fish my right contact out and as luck would have it, the contact rolled up into the upper part of my eye. This had happened before, so I went to the left eye and it came out as expected. Back to the right eye. I tried to fish the contact down to the proper position and finally did. When I went to take it out, it flew out of my hand and went onto the cobblestone sidewalk. The light was next to nothing and being an uneven sidewalk neither Steve nor I could see the contact. Paul and Joe went on without us. Steve and I stayed behind to look for my contact. After about five to ten minutes. I told Steve we should forget it and go on. We started out and after a few steps, Steve asked "you do have another pair at the hotel, don't you Dad?" I replied "No, this is the only pair I brought with me to England." At his insistence, we went back and started to look again.

Not too much later, three people, two women and a man came from across the street and walked right in front of me. The woman in the middle swooped down with out breaking stride and picked up my contact which was right in front of me. She placed it in my hand stating "Here is your contact, sir". I stood there dumbfounded, my mouth wide open, as they walked out of sight. I looked at Steve and he looked at me and we both had such a baffled look. This was crazy. When we caught up with our other men I asked if they had discussed losing my contact with anyone. They assured me they had not spoken about the contact to anyone. I tell you this story because I believe God was watching out for me and blessed me with the presence of a special angel that night. Several memorable things were happening on the day of that speed run that I would always remember.

After the Air Show in London was over, we were assigned to Mildenhall Air Force Base. This base is located in the northern part of England. There we would stage for our return flight home. On this flight home to Beale the SR-71 would be setting another record for flight from London, England to Los Angeles, CA.

We were tasked to fly home on the 12th of September. We started the launch procedures as normal and preflight started 24 hours prior to launch. The flight crew was connected to the aircraft on time and we launched on time. Things were going just great, but after taking on fuel and doing a test acceleration to speed, the flight crew reported a Red OIL Light. This is a "Land at the next base warning". The crew returned to Mildenhall and we did a post flight inspection and after de- briefing the crew, the engine shop found a wiring problem in the engine compartment, which they quickly fixed. The next day, Friday the 13th the SR-71 flew back to the United States setting a new speed record.

Chapter VIII—Coming Home

The return home was another chapter in new starts. The Mildenhall Air Base was to be our new overseas location called DET IV. We were generating plans to send a detachment to fly out of there but the English government baulked and did not want to upset the oil rich country's that we were wanting to spy on. After several months of dealing and arm twisting, we were able to locate an SR-71 there along with personnel to operate the task. We did this short term to begin with and then it became routine.

The Lockheed office at Beale AFB was the supplier of technical advice to the 9th SRW. We now went along with the aircraft to England, as well as Okinawa, sharing a three month rotation. We rotated in and out of Okinawa for over twenty years, and more than fifteen years out of Mildenhall. It was incumbent upon our Lockheed representatives to cover the airplane at both these bases as well as any area where the plane landed away from its home station.

My job was now on a rotation between Okinawa and England, typically three months at home, three months at Okinawa, three months home, three months at England. It wasn't long before Lockheed bid for the complete maintenance coverage of the SR's in England . We had civilian maintenance coverage at England rather than military.

As the SR became less secret, we were allowed to bring our wives with us on deployment. Most of us opted to take advantage of this opportunity and have the time to show our wives just what we had been doing on all these overseas trips.

I could not bring Pearlene with me at first as we had small children still in school. Later however, I was able to bring her on two occasions. Sadly, Pearlene was diagnosed with colon cancer in 1983 and although we went through all of the recommended treatment available at the time, she lost her battle with cancer in June 1987.

The SR-71 was delivered to Beale in 1966 and the flight crews were assembled. They were trained in every aspect of the flight regime and where they would be flying. We had a simulator that was as modernistic as the SR-71. In a space of two years, we were operational and replaced the CIA A-12; Black Shield also known as Oxcart. We never lost an SR to enemy fire of any kind and we never lost a life in an SR. We set several world speed records, altitude records not to mention the other trophies awarded to us by various organizations.

With the Speed Run of 1974, I was allowed to represent the SkunkWorks with the most futuristic aircraft the world had ever known. On September 1, 1974, we set the world's record for speed from New York to London....ONE HOUR AND FIFTY FOUR MINUTES!! On the return trip, we set a world speed record from London to Los Angeles on September 13th...THREE HOURS AND FORTY SEVEN MINUTES!! On July 27, 1976 the SR-71 set the world speed record of over 2092 miles per hour.

After the New York to London mission the SR-71 was seen at every prestigious air show you could think of. As always, we followed our "girl" as representatives. We were scheduled to go along with the SR-71 and be ready to aid her in any way needed.

We had a change of supervision at Beale . Paul Mellinger had been under pressure to retire and he finally gave in. There was also a rumor going around that the program was coming to an end.

Fred Carmody had been placed in charge of the SR-71 and U-2 programs at Beale. Fred was very straight about telling us that there were changes coming. I was told that I would be on the second round of layoffs in the late 1980s. However, this layoff never happened. I was able to send the last airplane out of Okinawa, Japan in January 1990 as the program ended.

This program was the most exciting program that I have ever been associated with and this is the statement from practically all of the personnel that I have ever know that was in this program. The collection of people, both civilian and military, were the most phenomenal ever and may never be repeatable. This is my opinion, but we did have the BEST OF THE BEST.

Much to my regret the SR-71 was retired from active duty in January of 1990. On the last flight of one of the SR's, she set several speed records that are still on record today. It will take a great air craft to top what she was able to do even at the end of her career.

These SR-71 airplanes are on exhibit at several air museums around the United States. If you can visit one of these planes at a location where they are displayed, I think you will be as thrilled as I am to be around her. In some of these museums there are symposiums where you can meet the pilots, RSO'S, maintenance personal and a tech rep (me), if you so desire.

Just like a beautiful woman, she had a sweet aroma. Her aromas still linger in my mind; JP-7 jet fuel and TEB (Tri-ethel borate). This chemical was used to ignite the fuel of the SR. I loved everything about this airplane!

I had been scheduled to go to Okinawa in late November of 1989. I took my new bride Nancy with me. We had made an earlier trip in January of 1988 where we spent our honeymoon. On the trip in 1989, we were told that the program was shutting down and I would have the responsibility to close down our office. This included shredding many of the secret and top secret, classified material. I had to mail back any data that needed to be sent back to the Lockheed factory. It was a bitter sweet pill but it had to be done.

Twenty plus years at Okinawa and fifteen years at Mildenhall were coming to an end but we had to get our "girl" ready for any possible missions that might be needed before we closed up shop. I was there for the last flight out of Kadena. We arrived at the hangar early and picked up on the count down to launch. I had one of our mechanics ask what I thought about this last flight. I replied; "This will be just as exciting as the first one….just going home".

Chapter IX The end is near or maybe not…..

Back in 1995, there was an actual attempt at a revival for three of the SR-71s. Three were fully refurbished for flight and loaded with new sensors and new necessary equipment. This one shot deal from Congress reactivated the aircraft and Congress also gave them a $100 million dollar check to go with it. The Advanced Synthetic Aperture Radar (ASARS) had been completed but not yet fully field tested. This was brought in with the reactivation. There was also a new Common Data Link (CDL) which brought real time imagery to the SR-71capabilties. There were other improvements like a clip in kit to replace the three vans that were the Operational Deployable Processing System used to process the ASARS data. The ASARS-1 was an all around radar sensor in the nose section of the SR-71. It allowed for the CVDL to be downloaded to a receiving station. If the system did go off line for some reason, the information could be held in the aircraft until it got close to another download station. . One other advantage was the Optical Bar Camera, which was a panoramic camera in the nose of the aircraft that was able to cover a 70 mile swath in one pass. It could use wet film and had the same type of processing as the U-2 used. The new Technical Objective Camera (TEOC) was a point and shoot framing camera which was also a wet process camera. New electronic countermeasure systems (ECM), DefA2, Def H and Def C were all brand new in 1995. These were some of the newer "toys" that the reactivation brought with it in the hopes of keeping this great aircraft alive.

Even with all of this, it was not to be. The SR-71 was cut from the budget in 1999. With all the new accoutrements, it just wasn't enough to keep the system alive. The aircraft were sold to NASA to be used as high speed test beds for the "Aerospike" program. The SR-71 in active duty was no more. Now, she worked for NASA at Edwards AFB and the Dryden Flight Research Center.

Through the efforts of many of the men that were on the program, there are symposiums at different locations that allowed the former flight team members to discuss their experiences. These people were from both enlisted and officer ranks. I have had the privilege to sit on more than one of these panels and looked forward to greeting many friends who would like to come and hear our story.

Chapter X This is not the End.

There is only one way to close this story of my journey. This is not the end. As long as the SR-71 stands in museums or airparks, as long as there is a hunger for the public to learn about this amazing piece of engineering, as long as there is someone that can appreciate the beauty of the Lockheed Blackbirds, all of them....the A-12 OXCART, the YF-12 Interceptor, the MD-21 the A-12 that carried the mach 3 D-21 Drone and of course, the magnificent SR-71... she will live! The legacy and legend of this family of aircraft, especially the SR-71 *my girl,* will continue to be written about by those who appreciate the majesty of this program and the brilliance of Kelly Johnson and all my compatriots that worked with me on keeping her up and flying. This was the highlight of my working life and I was compelled to let my story live as part of the Blackbird legacy. Many years of my life were given to this "girl". I missed many of my kid's birthdays, other family events, lived through the tragedy of losing my wife Pearlene and the joy of finding Nancy. The SR-71 was part of my life as sure as the blood in my veins. That is why I say.... This is not an end. My story is just part of the dark puzzle that is the life of the Lockheed Blackbirds.

THE SR-71 LIVES ON IN THE HEARTS OF ALL WHO WERE
PART OF THE PROGRAM
And in the hearts of all those who look to the skies.

Appendices

A-12 #121 the first of the A-12s getting some chine work at the Area 51 hangar. (Phoenix Aviation Research Collection/Lockheed)

The Lockheed test aircraft –SR-71 #972 with the Lockheed Logo on her tail (Seymore the Skunk) with shock diamonds trailing (Phoenix Aviation Research Collection —Lockheed)

A-12s #926 and #927 on construction at the Lockheed Burbank Plant (Phoenix Aviation Research Collection/ Lockheed)

Evolution of A-12

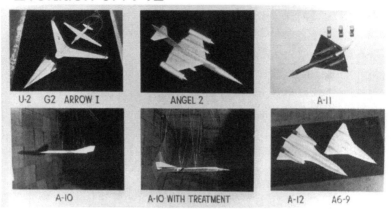

U-2 G2 ARROW I ANGEL 2 A-11

A-10 A-10 WITH TREATMENT A-12 A6-9

The Evolution of the A-12
(Lockheed)

BEN RICH SPEECH

SR-71 RETIREMENT CEREMONY

BEALE AIR FORCE BASE

26 JANUARY 1990

RECEIVED VERBAL APPROVAL OF SPEECH FROM COL. B. HENNESSEY
AND WRITTEN APPROVAL FROM COL. W. GRIMES
ON 25 JANUARY 1990

The following pages contain the retirement speech given by Ben Rich of
Lockheed
For the SR-71 retirement Ceremony

CONGRESSMAN HERGER, GEN. CHAIN, HONORED GUESTS -

TONIGHT I SALUTE THE MEN AND WOMEN OF THE 9TH
STRATEGIC RECONNAISSANCE WING, THE FLYING CREW, THE
MAINTENANCE PERSONNEL, THE PLANNERS, ANALYSTS AND ALL
THOSE ASSOCIATED WITH THE SR-71'S - AND LET'S NOT
FORGET THE SUPPORT TANKERS AND THEIR CREWS AND THOSE
FROM AFLC, FROM NORTON AND PALMDALE, WHO GAVE US ALL
THE LOGISTIC AND FLIGHT TEST SUPPORT. ON BEHALF OF
THE LOCKHEED CORPORATION, AND IN PARTICULAR THE MEN
AND WOMEN OF THE SKUNK WORKS, IT IS AN HONOR TO BE
HERE. I ALSO WANT TO THANK THE MEN AND WOMEN OF PRATT
& WHITNEY WHO DEVELOPED THE MARVELOUS J-58 BLEED BY-
PASS TURBO JET ENGINE.

I CANNOT HONESTLY SAY THIS IS A HAPPY DAY FOR ME,
SEEING THE RETIREMENT OF THE SR-71'S - I CAN SAY I
HAVE SEEN THEM FROM CRADLE TO GRAVE.

I REGRET THAT TWO OF MY FAVORITE PEOPLE CANNOT BE
HERE: KELLY JOHNSON, FATHER OF THE SR-71, SERIOUSLY
ILL IN THE HOSPITAL, AND THE LATE GEN. JERRY O'MALLEY.

THIS AMAZING BLACKBIRD IS THE FRUIT OF THE GREAT
MILITARY-INDUSTRIAL COMPLEX, THE U.S. AIR FORCE AND
THE LOCKHEED SKUNK WORKS.

- 1 -

52

I ACKNOWLEDGE AND ACCEPT THE NEED TO RETIRE THE SR-71
BECAUSE OF THE BUDGET SQUEEZE - BUT I DON'T AGREE
WITH IT.

I AGREE IT IS EXPENSIVE, BUT SO IS THE FIRE
DEPARTMENT, THE POLICE DEPARTMENT AND LIFE INSURANCE,
BUT THAT DOESN'T MEAN YOU GET RID OF IT.

NOW LET ME GIVE YOU SOME "GEE WHIZ" FACTS ABOUT THIS
BEAUTIFUL, EXOTIC, ONE OF A KIND, AMAZING FLYING
MACHINE - THE SR-71. I CANNOT GIVE YOU PRECISE
NUMBERS BECAUSE THE SECURITY FOLKS WILL HAVE A
CONNIPTION!

1. THIS AMAZING MACHINE HAD ITS FIRST FLIGHT ON
   DECEMBER 22, 1964 WITH BOB GILLILAND, OVER 25
   YEARS AGO. BOB EXPRESSES HIS REGRETS HE COULDN'T
   BE HERE TODAY, AS HE HAD TO BE IN ALASKA AND
   COULDN'T GET OUT OF IT.

2. WE DELIVERED THE AIRPLANE 12-1/2 MONTHS LATER TO
   BEALE ON JANUARY 7, 1966, - 24 YEARS AGO.

3. WE WILL RETIRE THE AIRPLANE WITH ALL ITS SPEED
   AND ALTITUDE RECORDS INTACT.

- 2 -

4. IT IS THE ONLY COMBAT AIRPLANE IN AIR FORCE
   HISTORY TO RETIRE WITHOUT THE LOSS OF A SINGLE
   CREW MEMBER - ISN.'T THAT INCREDIBLE FOR SUCH A
   SOPHISTICATED AIRPLANE - NOT ONE AIR FORCE PERSON
   LOST HIS LIFE IN WAR OR PEACE WITH THIS AIRPLANE!

5. ONE OF THE FEW AIRCRAFT THAT FLEW IN A COMBAT
   ENVIRONMENT AND WAS NEVER SHOT DOWN.

6. THIS IS THE FIRST TIME SINCE 1968 - THAT ALL THE
   AIRCRAFT AND ALL THE CREWS ARE IN THE U.S. - 22
   YEARS OF CONTINUOUS OVERSEAS DUTY. IT WAS AN
   AIRCRAFT DESIGNED FOR PEACE NOT FOR WAR. IT NEVER
   CARRIED ANY WEAPONS.

7. THIS MACHINE HAS FLOWN ALMOST 65 MILLION MILES,
   HALF OF THEM OVER MACH 3 - THAT IS EQUIVALENT TO
   2,600 TRIPS AROUND THE EARTH OR 135 ROUND TRIPS TO
   THE MOON OR TWO ROUND TRIPS TO VENUS.

8. I REMEMBER ONE FLIGHT FROM SAN DIEGO, CALIFORNIA
   TO SAVANNAH BEACH, GEORGIA IN 69 MINUTES!

9. ON NUMEROUS OCCASIONS IT FLEW HALF WAY AROUND THE
   WORLD AND RETURNED - FROM THE U.S. TO MIDDLE EAST
   AND BACK; FROM ENGLAND TO LEBANON AND BACK; FROM
   OKINAWA TO THE PERSIAN GULF AND BACK.

- 3 -

10. IT IS THE FIRST OPERATIONAL STEALTH AIRPLANE.
    ITS RADAR CROSS SECTION IS WHAT THE B-1B IS GOING
    TO GET THIS YEAR.- WE HAD IT 25 YEARS AGO!

11. IT WAS THE FIRST AIRCRAFT WITH STRUCTURAL
    COMPOSITE STRUCTURE. A COMPOSITE STRUCTURE THAT
    WAS CAPABLE OF 800°F - THE TEMPERATURE OF A
    SOLDERING IRON.

12. THE AVERAGE SURFACE TEMPERATURE OF THE AIRPLANE IS
    AT 550°F. FOR THE COOKS IN THE AUDIENCE - THAT IS
    THE TEMPERATURE OF THE BROILER IN YOUR OVEN. CAN
    YOU IMAGINE HOW SHOCKED I WAS WHEN THE AIR FORCE
    WANTED THE RED, WHITE AND BLUE STAR & BAR ETC.
    PUT ON THIS AIRPLANE! SOMEDAY PAINT A METAL PLATE
    AND PUT IT UNDER YOUR BROILER. IT'S HARD ENOUGH
    TO KEEP THE PAINT ON, BUT KEEP WHITE - WHITE, IT
    TURNS TO GRAY; BLUE TURNS TO PURPLE AND RED TURNS
    TO MAROON. WE FINALLY DID IT, IT WASN'T EASY.

13. IT IS THE ONLY 20 YEAR PLUS AIR FORCE AIRPLANE
    THAT NEVER HAD WING CRACKS OR NEEDED ITS WINGS
    REPLACED SINCE IT IS MOSTLY TITANIUM.

14. WHEN THE AIR FORCE WANTED TO SIMULATE HIGH SPEED
    RUSSIAN FIGHTERS SUCH AS THE MIG23 FOR SUPERSONIC
    INTERCEPT MANEUVERS, THE SR-71 HAD TO SLOW DOWN.

- 4 -

15. KELLY OFFERED $100 TO ANYONE WHO COULD SAVE TEN
    POUNDS ON THE AIRPLANE - NO ONE COLLECTED. I
    SUGGESTED USING HELIUM IN THE TIRES INSTEAD OF
    NITROGEN OR AIR, BUT HELIUM WOULD JUST LEAK
    THROUGH THE RUBBER. I SUGGESTED TO GIVE EVERY
    PILOT AN ENEMA BEFORE EVERY FLIGHT - THAT DIDN'T
    GO OVER VERY WELL!

16. IT'S THE ONLY AIRPLANE WHERE THE HYDRAULIC OIL
    COST MORE THAN SCOTCH WHISKEY!

17. THE POWER PLANTS HAVE THE THRUST EQUIVALENT TO
    THAT OF THE QUEEN MARY.

I COULD GO ON AND ON.

LET ME TELL YOU ABOUT ITS BIRTH - - -

KELLY GATHERED ABOUT 75 PEOPLE TO DEVELOP THIS
AIRPLANE - BUT THE CONFIGURATION WAS PUT TOGETHER BY
FIVE OF US - FOUR OTHER ENGINEERS AND MYSELF, DRAWN ON
A SPARE DOOR LAID ACROSS TWO DESKS - THAT WAS THE
BIRTH OF THIS MAGNIFICENT MACHINE.

- 5 -

WE CORRUGATED THE WING SKIN, SO THAT AS THE SKIN
EXPANDED - ANY WRINKLES WOULD BE STREAMWISE AND NOT
CAUSE DRAG.

THE AIRPLANE GETS 65% OF ITS PROPULSIVE THRUST FROM
THE INLET; 25% FROM THE ENGINE; AND 15% FROM THE
EJECTOR NOZZLE.

I REMEMBER CALLING THE J-58 ENGINE THE MACY ENGINE -
BECAUSE THEY SPENT SO MUCH MONEY. I TOLD BILL BROWN
OF PRATT & WHITNEY THAT IF I GAVE THAT MUCH TO MACY'S
- THEY WOULD HAVE GIVEN ME THE ENGINE.

THAT'S ENOUGH POT-POURRI FOR TONIGHT. I WANT TO THANK
EACH AND EVERY ONE OF YOU HERE TONIGHT WHO WAS
ASSOCIATED WITH BLACKBIRD - THIS NATION OWES YOU ALL A
GREAT TRIBUTE.

I'LL CLOSE BY SAYING - YOU DUN GOOD!!

- 6 -

NOTES

NOTES

NOTES

NOTES

NOTES

Printed in Great Britain
by Amazon